"The bells that toll for mankind are like the bells

of Alpine cattle.

They are attached to our own necks, and it must

be our fault if they do not make a tuneful and

melodious sound."

Peter Medawar

Our amazing world

Seen by a scientist, a thinker, an Astronomer Royal

Martin Rees

edited and adapted from his writings

by Ruth Finnegan

OUR AMAZING WORLD

ISBN 978-1-911221-57-9

HEARING OTHERS' VOICES

Balestier Press

Singapore and London

www.balestier.com

CONTENTS

Join our community. 167

Beginnings

I was born in 1942, in York, England.I was born there because my father was doing his basic Air Force training at some camp near York and so, by coincidence, we had to be there for the year. But my family, my real upbringing, was in the Welsh border country.

This was because when I was one, my mother moved back to North Wales. And then I was afterwards brought up mainly in Shropshire, just on the borders of Wales.

My parents were school teachers, They met when they were both teaching at the same private school in the late 1930s. My father was a

Cambridge graduate in English, and taught in secondary schools and at Trinity College, and my mother was trained as a primary school teacher. She taught in a variety of schools. And in fact, just after World War Two, she briefly ran a small private school herself.

My earliest firm memories really date from 1948 because in that year, my parents decided rather boldly that they would start their own school, a progressive boarding school. They rented a large somewhat rundown mansion in South Shropshire. We moved there, and they opened this school which is still going. So my early schooling and my early life were in a very small village in these really beautiful surroundings on the hills of South Shropshire. Itwas deep in the country, with Ludlow as the nearest market town, part of my childhood , a beautiful small town.

As an only child I got a great deal of attention. Particularly in my first few years, when my mother was looking after me, I must have had a tremendous amount of personal attention. And my parents were extremely supportive. I obviously derived far more education from them than from anyone else for my first 12 years of life. My father in particular was an extremely versatile, talented man who painted, composed and represented in politics, etc. And so I think I learned a great deal from him, as well as from the formal education I had at that time.

Up to the age of 13 I went to the school my parents ran but at 13, as for so many other children at that time, there was a choice. I could either go to the nearest grammar school, a 14mile bus ride away, or be sent as a boarder to Strawsberry

School, 30 miles away. My parents chose the latter. So from 13 I was at that boarding school.

I don't think I had any particular interest in astronomy or in science. I was just into figures and numbers from an early age, and into how things worked. But I think I was just as interested in literal things and in history, and if I look at how I ended up being a specialist in science, it's not through any early commitment or ambition. It's really that during my school career, I always found languages the worst subjects, and so it was really languages, I tended to shift away from when specialization was needed, and into science and mathematics.

I was very lucky. It had a rather old fashioned public school ethos, which even then I didn't really relate to optimally or I respected much, but it was extremely well staffed, especially in sciences. I was very well taught in all subjects, particularly in science.

I recall, in particular, the senior physics master at that time, who was about 60. He was a Cambridge graduate from, I suppose, the 1920s. He had been told, when it came to deciding in a career that he wasn't good enough to do research, because there were better people in his year.And so people of my generation were the beneficiaries of outstanding teaching by committed and very able people.

His name was Bill Matthews. And he was one of several excellent teachers that I had. Most of them are no longer with us but one who is still alive, Jeff Chew, went on to be the Keeper of the Science Museum. He was still alive in his 90s and I remember very well his magnificent teaching of physics. When we needed notes he produced them on some primitive duplicating machine for the entire class.i so remember the very good stuff that he gave to all his pupils in the science.

I am extremely grateful that due to that I did well enough to get myself into Cambridge.

I liked music too. I have no great talent for it, but I was put to the piano when I was a child for which I'm very grateful (I like to play slow movements when no one's listening!). I even sang in the choir, something for which I'm again grateful, and for my memories of the school altogether.

One of my best memories is of the chapel services. I've never been religious but certainly the singing in the choir in the congregation was something that was one of the most pleasant features of a school like that.

The high prestige subject in the school was classics, which I didn't do, but fortunately they also had, as I said, very good teaching of maths and science. I specialized in Mathematics and Physics, the subjects I did for A level.

Then when I was told to try for Cambridge I applied to Trinity College, and got in - a good choice.

When I got there i studied mathematics, though in retrospect, I might have done better to have studied science. In Cambridge, you do a range of subjects for broader education, I did the mathematics tripos. My family generally are all good at maths, although I knew quite early in my education I wasn't cut out to be a *real* mathematician, which some of my contemporaries were. I enjoyed the problem solving aspects of maths , which I was quite good at. But I wanted to do something else, so in my final year in Cambridge, I shifted away from pure maths towards the *applications* of maths.

I didn't know what to do, really. At one stage I thought I would do economics as I had some friends who'd shifted from math to economics. I thought about that quite seriously.

But I did manage, just by bit of luck actually, to get a research studentship to stay in Cambridge. And my most important piece of good luck was being assigned to a very good supervisor called Dennis Sciama

When I started my PhD project, I had no idea whether I would be suited to it or whether I would enjoy the subject at all. But within a year, he had got me enthusiastic and I was convinced it was something I was going to be happy to be doing, at least for a PhD. And I was lucky at that stage, partly in having this very good supervisor, but partly also in the choice and timing of my research topic, it was 1964/65 when I started as a researcher in the subject of astronomy. Cosmology was then at an important stage because for the very first time there was discussion about whether black holes existed, whether we were thinking about the Big Bang, and so on.

So it was possible to follow developments, which were, in retrospect, very important in the subject. And of course I was also working in a subject where everything was new. When you're young, as I was, it's possible to immediately start making contributions that are actually on a level with those who are more experienced. And so, in that sense, I was very lucky to start thinking about topics that were new, and where it was possible for a young person to quite quickly make original contributions. It helped too that that was the time when Radio Astronomy was just beginning to really make a mark, with Cambridge a big center for that (more on that later).

When I passed my PhD, I got a research fellowship, first at Jesus college, and then almost immediately went off to the United States for short spells. I was so lucky to get that fellowship. And then I did a sandwich arrangement, a good idea because it was about four to six months in California at Caltech, then back, then six months in Princeton, then back here again. Wonderful.

That meant I got some very good experience without feeling worried that I was going to be out of a job. So I was fortunate to have that opportunity. And then I had a base in Cambridge, which I could use a bit longer.

In 1972, I moved to Sussex University as a professor. At that time, I had no expectation that I'd come back to Cambridge at all. I remember throwing away my gown, thinking I'd never wear such a garment again! But as it turned out, this was the year before Fred Hoyle resigned unexpectedly from his chair in Cambridge, and I was appointed to succeed him.

So I came back, and I've been based in Cambridge in various capacities ever since 1973.Of course I've also made visits elsewhere, like to the States

I've been lucky in that for my subjects. Cambridge has been a strong center and I feel I've done my bit to help to keep it that way. I think we are still a major center for my subjects, and I certainly don't feel like I could have done my work any better anywhere else. Certainly having visited most of the major centers in the US, I haven't found anywhere Ithat offers the combination that Cambridge does of good developmental structure in nice surroundings (which we have at the Astronomy Institute), plus the extra dimension of colleges and interdisciplinary contact. And just the general ambience too and the community and the aesthetic background.

If you transplanted King's College chapel to California I might be slightly less reluctant to go but there's nowhere else I've met within the States that has that combination of attractiveness. So I've never felt attracted to the United States, let alone becoming a US citizen.

The opportunity to reach across to other disciplines and ithrr researchers has been important to me all my life, interacting with the community in our subjects. Astronomy is well known, and over many years, for a very interactive, informal atmosphere where everyone gets together for coffee, everyone works with   their doors open and everyone is in and out, with lots of visitors coming and going too, etc. That makes a stimulating atmosphere. And it's very important indeed, especially for young people, to be in an atmosphere where they meet senior people: you don't have to go and knock on their door, they're always just there for coffee etc. That aspect has been very important.

That was a decadein Cambridge when radio astronomy was flourishing .some of the first evidence against the famous steady state theory came from the radio astronomers in Cambridge. And also, the first evidence that black holes existed, in fact came from byproducts of the work they were doing.

It was an exciting time. And I found that it suited my style of thinking, in that I've never been all that good at long deductive chains of reasoning, I've always prefered more sort of synthetic or synoptic style of thinking. And so what I liked doing at that stage was trying to make sense of some fragmentary and disconnected seeming data, and seeing if it fits into a pattern so you can explain it. And that actually is the theme of most of the work I've done ever since trying to sort of make sense of data been close to the observations and the phenomena, trying to apply what I know of the theory to make sense of it.

Though I wasn't myself a member of the   radio astronomy group, I went to their seminars, and I knew the students who were my contemporaries like Jocelyn Bell who was in on the discovery of pulsars. And so I was very much involved in talking to them.

As you will see a little later.

The contact with people in other departments and other spheres has also been very important

for my sort of personal education and broaden-
ing. So I had the chance to get to know people
like Cynthia Brando, Frank Kermode, Dan McK-
enzie and people like that who were leaders in
their subjects.

To have that, even when you're a young person,
is wonderful, It means that even younger faculty
in the university are embedded in a real cross
disciplinary atmosphere.

I also found it helped me very much when I was
a young professor to be involved in, for instance,
selecting college research fellows. You're trying
to compare people across different subjects and
get a feel at an early stage for the different criter-
ia and the different standards and value systems
across the whole range of academic subjects. I
found that very helpful and very enlightening.

And so that's something which you routinely get all along here, whereas it's an experience that only senior people get in the US system.

I also liked the social and convivial atmosphere of Kings College that I was privileged to be part of for a long time. Actually I think that rather exemplifies what you do find in Cambridge. I would say Kings is unique - certainly, I was privileged to be there. And, of course, it is famous in the world for its chapel and I'd say that the excellence displayed by what goes on in the chapel is an important feature of the college,. One of my happiest experiences there was walking through the doors at night when someone was playing in the chapel. The whole thing is like some huge musical instrument, a wonderful experience that you don't get in many other places.

Then after some years I had the honour of being elected a Fellow of the Royal Society, and then, unexpectedly, its President.

The Royal Society, and more

On the 28th of November, 1660, following a lecture by Christopher Wren, group of dustnguished men such as   Samuel Pepys and Robert Boyle set up "a Colledge for the promoting of Physico-Mathematicall Experimentall Learning"

The group received royal approval, and from 1663 was known as 'The Royal Society of London for Improving Natural Knowledge'.

This then was the start of The Royal Society (https://royalsociety.org) an acclaimed national institution with international influence. It still functions as that today in its impressive building, a Fellowship of many of the world's most eminent scientists and the oldest scientific academy in continuous existence. It has had an intriguing history, the core of our modern-day scientific discoveries, and it has - rightly - been said that the story of the Royal Society is the story of modern science.

THE
ROYAL
SOCIETY
7
7

"'The indeoendent scientific academy of the UK and Commonwealth, dedicated to promoting ex-cellence in science".

The Royal Society's motto 'Nullius in verba' is taken to mean 'take nobody's word for it'. It is an expression of the determination of Fellows to withstand the domination of authority and to verify all statements by an appeal to facts determined by experiment.

Over the years many distinguished scientists have been elected as Fellows : Isaac Newton, one of the earliest Fellows and President 1703–1727

Lord Kelvin, Stephen Hawking, Fred Hoyle, Jocelyn Bell Burrell ... too many to,list here. At present it has a Fellowship of something around 1600 eminent scientists, and since its foundation over 8000 in all - again too many to name here (for a full list see https://en.m.wikipedia.org/wiki/List_of_Fellows_of_the_Royal_Society).

I'm also greatly honoured to hold the post of Astronomer Royal.

Today this is an honorary post. It means simply "one of Britain's most senior scientists". Like a Chief Scientific Advisor, or the head of a scientific society, the Astronomer Royal can be expected to give all sorts of opinions about science and science policy, straying at least occasionally, if they wish, well beyond their area.

Was it always like this? Yes and no. Until the 1970s the post of Astronomer Royal was synonymous with director of the Greenwich Observatory (at Greenwich, Herstmonceux and then Cambridge). Before the 19th century, the AR was also an active observer, in fact only one of two observers in the institution. Nevertheless, Astronomers Royal were often called upon to make judgements and offer advice in areas that did not relate to making observations or managing an observatory. Because the Royal Observatory was funded by government, being under the administration of first the Board of Ordnance and then the Admiralty, there was potential for ARs to be asked to consider a whole range of technical and scientific issues.

The first Astronomer Royal was John Flamsteed. The self-taught and highly accomplished Flamsteed earned £100 a year after Charles II issued a warrant to the Office of the Ordnance in March of 1675. In June 1675, an additional warrant allowed for a small observatory at Greenwich. Sir Christopher Wren, a keen astronomer in his right, was appointed to build an observatory. We know it today as The Royal Greenwich Observatory.

In 1710, Queen Anne issued a warrant naming the President of the Royal Society along with other Fellows to serve as Visitors to the Observatory to try and halt its decline.

Flamsteed had a bit of a problem publishing his work as quickly as all had hoped, so, due to this lack of published work, George I named Edmund Halley as successor to Flamsteed in 1720. Halley would become 'Our Astronomical Observator in Our Observatory at Greenwich.' Halley would later become famous for his prediction of his namesake comet.

In 1727, Queen Caroline became the first Royal to visit the observatory. Halley would then receive a supplement to his meager earnings as he received half-pay of a Post Captain.

During Nevil Maskelyne's term (1765-1811), it was discovered that The Royal Society warrant lapsed when Queen Anne died. Thus, the Society was without authority for 50 years. Since then, a new warrant is issued by each new Sovereign.

As many appointments have a set structure, the Astronomer Royal has never really formulated a set routine for appointments. The Queen or King makes the appointment with advice from the Prime Minister, who in turn has an official recommendation from the Board of Ordnance of Admiralty.

No duties are attached to the office. The Astronomer Royal though is expected to be available for consult on scientific issues for as long as the holder remains a professional astronomer in the UK.

For much of the AR's history, the most obvious place in which this happened was the Board of Longitude. While many of the ideas under consideration there were astronomical (involving knowledge of astronomical theory, mathematics, optics and instrumentation), others were based on geomagnetism or, of course, horology. Understanding clocks and timekeeping was essential to astronomy, but the specifics of horological theory and manufacture would have been beyond the AR's experience.

ARs also advised on areas like cartography, instrument design and weights and measures, that involved techniques closely allied to astronomy. But they were also asked to consider a wide range of fields of interest to the Admiralty and other branches of government, simply because they ended up being their available scientific expert.

One of the ARs who most obviously became the government's go-to scientific and technical guy was George Airy, who was in position from 1835 to 1881. Airy covered a great deal of ground, intellectually and practically. Unlike all his predecessors he was not much involved with daily observations and he had a significantly larger workforce at the Observatory, onto which observation, calculation and even management could be delegated. Airy, for example, did a considerable amount of work on the effect of iron ships' hulls on compass use and design. He also advised, like many other ARs, on education and he was involved in the organisation of the Great Exhibition. He was, perhaps most intriguingly, called in to advise the Great Western Railway on track gauges and the engineer Thomas Bouch about the pressures that might be exerted by wind on the planned rail bridge crossing the Forth.

That advice got him into trouble! It was first applied by Bouch to the Tay Bridge and, when that collapsed in 1879, Airy was called in by the enquiry. He claimed that his advice had been specific to the circumstances of the Forth and the design for that bridge (which was now speedily discarded). The enquiry agreed, suggesting that Bouch had "must have misunderstood the nature of [Airy's] report".

Airy did know quite a lot about engineering. He was, apart from anything else, closely involved with the design of large instruments and their mounts at Greenwich.

Times and the nature and range of expertise have changed considerably since the 19th century. I myself am not an Astronomer Royal who can offer specific or technical engineering expertise. What I focus on, now, is calling for research and funding.

So you can see that with all that when I became Astrinimer Royal and, for five years from 2005, President of the Royal Society I had a very illustrious lineage to live up to.

Actually I did rather enjoy being President of the Royal Sucuety. I was appointed to that office a year after I became Master of Trinity College in Cambridge   - another great privilege. Quite a time-consuming one it was too, meaning two or three days a week in London, and other opportunities. To be honest, if I had known I was going to take on Royal Society I probably would not have agreed to do Trinity was although they are both very privileged positions. It's hard to feel you're doing them both adequately, and have enough time for other things too.

How does the Royal Society compare to a Cambridge College? I would say the similarities outweigh the differences. The Royal Society is a larger and somewhat more formal structure but basically it's very much the same. The Chapel is not quite as spectacular although one thing is, we do have is outstanding music. The outstanding choir and in the chapel is important there too. So yes, I think the similarities greatly outweigh the differences.

I think the Master in Trinity is even more non-executive than the master of most other colleges in that there's a very large fellowship with a very, very strong commitment to the college. And one thing that impressed me hugely is the willingness of the fellows to devote time to college and is never difficulty finding people to be tutors and take on other worlds like that there's tremendous willingness to work for the college. And this of course, means that the the master is non executive and not so hands on practice in a smaller college, but certainly it's been a great privilege.

Back to the Royal Society - it has been changing a lot recently. By now it is very much concerned to look towards the future. In fact, one of the things that kept me especially busy in my role when I was president was that in 2010, we celebrated our 500th anniversary. And in connection with this, we were trying to expand our policy activities, international links, etc, and also play a wider role. But I think what makes the work fascinating is that it is engaged with policy over a wider range of subjects than before, essentially because some technology enters into a wider range of government policies, like the environmental energy, climate, and health agendas. Clearly science is very important in all of these.

And as the UK is part of pthe Commonwealth Academy of Sciences, we in our Society feel we want to maximize our impact on decision making to ensure that decisions are made on the basis of the best available evidence, so the international aspect of our work has expanded.

We have links obviously with Academies of other countries, the US and others. A lot of our work is on international issues like pandemics and climate change, etc. And one thing we started just two years ago was a routine meeting of the Academies of the G8 countries in the lead up to the G8 summit.

This happened first in 2005, when the UK had the presidency. The aim is to bring together the Academies of the G8 plus five others - India, China, Brazil, Mexico and South Africa. These try to academies to make unanimous decisions rather than platitudinous mistakes. I'd say it's important that you have this, and important that it feeds into the G8 summit.

.

That's an example of how activism broadened in the international context, and that's what the Royal Society has been trying to do. But I suppose above all it is aimed at trying to maintain scientific excellence and support the quality of science, particularly in this country. I think this is crucial for this country because we have an enviable tradition of scientific research and data dating right back to the founders of the Royal Society in the seventeenth century.

It's important, I believe, that we maintain a competitive edge in a world where the competition comes not just from US, but from the far east as well, relevant for maintaining the UK's prosperity. We have a head start in our traditions in science but we do we have to sustain them. Actually we're fortunate to have had not only positive rhetoric from successive guvernments but positive actions towards this end. I hope it remains that way.

And I hope the Royal Society is going to be able to have an increasing impact on this sort of thing and that it will be as broadly involved in activities inside of politics as our founders way back in 1660.

Science is worldwide, so I have now been to China, also to Japan and India, quite frequently. I think we've got to realize that in the 21st century the intellectual capital of the world is going to be concentrated in Asia, and not in Europe or the United States That's where the population is concentrated. And it's a highly educated population now, and so the balance of intellectual effort is probably going to shift from Europe and the United States towards Asia. We need to be mindful of that and adjust our policies accordingly.

Famous scientists I've worked with

The discovery of pulsars came in 1968. I remember it well. I was by then a recent graduate, spending a few months at Caltech as a visiting postdoc (post-doctoral student - after successfully completing a doctoral dissertation). I remember Dennis Sciama writing me a letter to tell me about this important discivery - quite different from emails we have today, instant communication. So I was the first person at Caltech to hear about this exciting discovery via that letter from Dennis Sciama.

Dennis Sciama was a fine role model. He was a very articulate person who had broad interests within and beyond science, and he was very good at being positive about one's work and give you encouragement. And he also have to have a good set of students. Hawking, Carter, Rob and a number of others, and we all learn from each other, as well as from him, and it was a lively atmosphere. So he was the most important influence. And Fred Hoyle was actually still around but at this stage Fred Hoyle was rather isolated figure because of an internal departmental disputes. And so it was Dennis Sciama who took al the students.

That's why Stephen Hawking and I and others were students of his, I only really got to know Fred Hoyle later because in 1967, which was the year I got my PhD, Fred Hoyle opened a separate Institute, which was separate for the math department and actually offered me a job as a postdoc in that department. And so I got to know him from then onwards, but as a scientific mentor, he was never as important as Sciama because although Fred Hoyle was unique figure of the subject and for 25 years from the 1940s to the early 60s, he was probably the person in the world who contributed most to physics in terms of original ideas. A real forthright Yorkshireman!

By the time I came into his orbit, he was becoming a slightly isolated and eccentric figure cut off from the mainstream. So, although I had a lot of social contact with him, he was never an intellectual mentor. And it was already clear that he was becoming rather isolated in his views from the mainstream because he was not interacting very much with the younger people.

He was an extremely warm character that I admired very much and kept up with until his death. In 1972, he resigned rather prematurely from Cambridge. I was by then a young professor at Sussex.i was appointed as a successor which brought me back to Cambridge. I'vealways appreciate the fact that Fred always supported me helped me in my career, even though I had been writing papers contradicting his preferred theory. I mean, he was never reconciled to the Big Bang. He believed in the steady state in later life and compromised and believed in some sort of steady bang.

He never really appreciated the kind of work that I and others were doing. But nonetheless, in personal terms, he was most supportive and I know that he was not unhappy that I had been appointed to the chair after he resigned.

And then Stephen Hawking. He was two years ahead of me so when I joined Dennis Sciama's research group, Stephen was in his third year. And of course, his disease had already started he was hobbling around on sticks at that stage, and Dennis Sciama told me that it was not [...] finished his PhD at the time that he was taking on research students. And it was astonishing, of course, that he did finish the PhD and will never predicted his future career. And, in fact, every sort of fast forward to recent times. I attended his 60th birthday conference and was more recent events 40 years later, and although as astronomers were used to very large numbers, fewer as large as the old time then have given back in the 1960s. He remained a friend until his death.

Over the years, in terms of work, our scientific interests slightly diverged, my own work has been more close to observations whereas he gradually moved away to more sort of speculative areas. And so in terms of research, we've never really been collaborators, but we always remained in contact. And his book on time had an immense effect. Apart from the work of Carl Sagan, with an even greater worldwide impact, Hawking made the widest impact of all professional scientists involved in astronomy and cosmology with the result that many people realized the subjects were of interest.

How science works - me too

Einstein's famous remark that "Unless an idea starts off as absurd, there's no hope for it" Is a great comment that makes a lot of people look very puzzled.

 "That's not how science works", they'll say, "*we* work on known facts. Maybe that's how mathematicians and physicists work but not *us*".

On the other hand Sydney Brenner, the Nobel biologist, immediately said "Oh yes, most of my ideas - I've written about this - start off as half right and half wrong. And then gradually, through conversations and evidence, I improve them until they work, but yes, many of them *are* pretty crazy"

True enough, most ideas do start off being rather fragile and vulnerable. So you have to protect them a bit before you jettison them. But I think that we also sometimes get a wrong impression, influenced by Thomas Kuhn, who emphasized the idea of science proceeding by normal routine, the occasional revolution. I think there've only been one or two revolutions in the Kuhnian sense in the whole history of physical science.

Most new ideas in science don't overthrow the old ideas. They transcend and generalize. For instance, Einstein didn't really overthrow Newton's laws, he extended them to apply more widely. In almost all the cases I know about the advances in science have transcended and subsumed what went before, very seldom does one really have to wipe the slate completely clean and start again.

So, science is really an evolutionary procedure and in most cases, in the subject I've worked on, it's been led not so much by brilliant ideas, but by new observations which, in turn, are made possible by technical advances. And certainly, in my subject, 95% of the credit should go to those who develop new technologies and new kinds of observation for experiments to be carried out, and it's *that* that leads to new ideas, new thoughts, new things to interpret.

And that's why young people don't have to batter their brains up trying to solve what the rest failed to solve, they can think about new things, new techniques.

So I think that is the nature of ideas that they evolve and develop. You try to fit some pieces of the jigsaw, to generalize rather than overthrowing everything and starting again. Of course, there are key insights and the most important developments in science do involve some new idea - a new unifying idea which links together lots of unrelated thoughts.

Of course, the other big contrast between science and humanities as regards creativity is that, although in science, your work is durable, it can lose its identity, you merely add your bit to the corpus of public knowledge.

Einstein is perhaps the only 20th century scientist who really made a distinctive imprint in science. Had he not existed some of those ideas might not have p emerged for decade. But mostly individuals play only a marginal role.

One of the most memorable statements about creativity in science and the humanities is in one of Peter Van Inwagen's essays, when he points out that when Wagner took 10 years off in the middle of the Ring cycle to compose the *Meister-singers*,. He wasn't concerned that someone would get there first. But if you are a scientist and you've taken 10 years off, someone else might have found the double helix!

In science, very rarely does any individual advance the subject more than a few years compared to what could have been had they not existed. But that doesn't diminish the satisfaction you get, if you feel you have had an idea that has been vindicated.

There *have* been revolutions in science. One was by Gallileo. Copernicus   too. And I think probably another one is the quantum revolution in the early 20th century because obviously, the concepts of quantum theory were completely counterintuitive. And though they were of course completely different from the mechanistic view, we do now realize that in a way they are consistent with it. It is only when we get down to the micro world that these counterintuitive, spooky, elements of quantum theory come in.

And, incidentally, that really leads to one of the other things that I marvel about when I do my own science, which is how far we can actually get with common sense intuitions. It is a remarkable fact that our brains, when they come to deal with everyday experience, can go so very very far, and interpret phenomena, far, far beyond the experience of our ancestors.

But we shouldn't be surprised that you can't extrapolate indefinitely. When you get down to the world of the quantum, then you seem to lose the everyday concepts of mechanics. And suddenly when you get up to the realm of the entire universe, again ideas of space and time that we've grown up with come in. I think the remarkable thing is not that we have to develop new intuitions for the extremes of the very large and the very small, but that our brains and our intuitions have been so successful in making sense of the physical world.

So, to get to the nitty gritty, how do I myself work?

Well, I sit and think. And often, of course, I just sit!

But I have also benefited a great deal from working in collaboration. And that's not just for the cynical reason that the other guy does most of the work (though I'm afraid that's true more often than not), but that it's very important to have someone with whom you can discuss the technical details of some issue in a way that the average person can't. If you're are working in collaboration, then not only do you have someone with complementary expertise, but you have also have someone with whom you can discuss the details.

Actually I do find that talking to people about ideas is something I enjoy. And a lot of ideas do develop in that sort of way through interaction, tossing ideas off one another. I think what I'm saying is quite a common experience: most artists, say, need to have a sounding board as it were.

So working on various topics I've benefited from a succession of collaborators, colleagues and students, etc. on. And obviously, sometimes they bring special expertise. They've often spent a lot of time talking to people who've made the observations and tried to make sense of them.

I think that the *social* - people - side of science is a very important part of it. When I go to conferences I think what's important is not only meeting the people, but also being able to calibrate how seriously you should take them! To tell the truth I've saved a lot of time by hearing some guy speak at a conference about work that I hadn't been able to fully understand and realizing, having heard him,   that it was not *my* fault that I couldn't understand him! So sometimes, from getting a view of the community where certain persons work in a certain style of work, you can decide how you could better use your time.

I find that I haven't wasted my days when I go to a conference as you can get a feeling for what the consensus is, the way we work now is very much affected by email and the internet. My work with collaborators is quite often by email, etc. But I think that only works if you've already met the people in the real realities of work. And *then* you can have a team collaboration. First, though, a real meeting is crucial, at a conference or wherever.

That's quite reassuring for me because having spent a lot of my life in Cambridge - one of the main centers in the world for this type of research

- I'd be rather sad if the benefits of it depended on sitting on email all the time. There is a tendency for that to happen of course, I find lots of my colleagues are on email, collaborating with someone else on a different continent. I think that real contact is still very important.

My own work has been in what I call cosmology. Although I'm a theorist trying to make sense of data and develop ideas, I keep in very close contact with *observations*. And I think the distinctive contributions I've made have been through trying to link together unrelated data obtained by different techniques, maybe optical telescopes, radio telescopes, spacecraft, etc, and trying to see if they fit into a pattern. So I've always been in close touch with data, with trying to make sense of them by applying the physics I understand, it's like, if you're an engineer, and you're told to use your knowledge to construct something to meet certain specifications, you do your best and sometimes you succeed, sometimes you fail. So what I'm doing is trying to use the laws of nature to understand and to make sense of some phenomena. And I've done this through most of my career since back when I started off.

At that time the first evidence was coming for the Big Bang origin of our universe and for black holes, etc. I was very lucky in that that was a good time to be starting as a research student - when the world could use new ideas. And what's been very fortunate is that my subject has not stagnated. And that, again, largely through the continuing development of technology, the rate of discoveries has remained very high. And if I look at what's happened in the most recent five years, then the pace of discovery has been at least as high as it was in the late 1960s when I began.

Science at work!

... and tries to communicate

We also need to thstart no about popular science writing. The details of our subjects are rather arcane, and technical but even so it *is* possible, I think, to put over the concepts to a new generation without too much elaboration. and without too much distortion either. Yes, I think it is possible to convey the ideas devoid of too many technicalities in a way that can be understood by non-specialists.

Actually, it's gratifying that the topic is found fascinating by so many people. I would certainly derive less satisfaction from my own work. If I felt it was only something I could discuss with a few colleagues.

If you look at what the wide public is interested in, they focus on the fundamental questions of origins: the origin of life, of consciousness, of the universe. And so these are the topics on which there's the largest flow of genuinely scientific popular books.

And I think it's a real benefit that in subjects like mine that have no direct practical relevance, it goes through to this wider public, and that work becomes somehow part of our everyday culture - seeking to understand something of the broad complex of the universe and the amazing chain of events whereby from some mysterious beginning 14 billion years ago, atoms, stars, planets, biospheres and eventually human brains have all evolved.

I'm glad to be one of the people to try and convey these ideas to a wide audience. I think it's good for professional scientists, whatever field they are in, to speak to people outside their special expertise. The reason for that is that if you're a scientist, then of course, it's part of your professional skill to focus on bite-sized problems that you think you can solve. And that often means of course, the focusing on what might seem trifling problems, because the big problem can only be solved piecemeal, it's always premature to go for frontal attack.

The occupational risk, therefore, of being a scientist is that you become so engrossed in the fascination of your specific minor problem, that you forget the big picture. And you forget that your problem is only worth solving insofar as it helps to illuminate the big picture. And when you talk to a general audience, then the questions they ask are the big questions, and that reminds you that those *big* questions are the important bonds and makes you see your own minor contributions to the overall pattern in a true proportion.

And after all didn't Einstein say "you should make things as simple as possible, but not simpler?"

Yes, because there are some concepts which are very hard to appreciate without the mathematical language that many people don't have.

A mistake many writers make when they write about science is to assume that by eschewing mathematics, you make it accessible. But you don't, because there may be unfamiliar vocabulary - well maybe familiar words being used in specialized senses, which can be very confusing to some readers. So I think one has to work very hard to present the work in a way that will be understood by someone who doesn't already have that vocabulary or those concepts.

Also, sometimes people use metaphors which seem rather clever to the insiders, but which don't really make sense unless you really know the stuff yourself. And so there are all kinds of pitfalls. And that makes me admire very much those people who do nonetheless do a successful job and convey the essence of science to another generation. I also admire journalists. They, of course, don't write careful books, but I know how hard it is to explain in simple language, even something I think I understand fairly well. And journalists have to describe something which may be unfamiliar to them, and do this to a tight deadline - a very difficult art. So I think scientists should not be in any sense snooty about those who do the popular writing and interpreting.

I find when I'm writing, for general audiences, it's very helpful to give one or two of the members of that class a draft so that I can see which bits they can't understand. Actually most of my writing and all my books have really stemmed from lecture notes or articles. I'm not a natural writer, I would never sit down from scratch and write a book with a blank sheet of paper in front of me. So what's happened is that I have done quite a lot of general lecturing. I've written quite a few articles. And having done that, I felt it worth the extra investment of trying to synthesize these into a coherent story in hardcovers. So my books have not been things where I start in the beginning and finish in the end but a way of seeing if I can merge together some earlier writings or fragments in order not to waste them!

To tell the truth I don't like lecturing, I don't like writing. The one thing I *do* like is *preparing* lectures. I like trying to think of how ideas knit together, and how you can best explain them. That means that I enjoy writing notes for the next thing that I have to give. I don't enjoy *giving* the lecture, it's actually doing the preparation and seeing how the ideas fit together. And I also enjoy informal discussions about my work.

Additionally I think one thing we all have to learn and many academics are rather bad at, Is not to be too much of a perfectionist. We all know that if we devoted more time, we could make our lectures better, polish our papers to make the them better, etc. But we've got to learn to maximize the *sum* of several things together and not try to maximize each. Some people are ultra perfectionist and spend too much time on one thing, and therefore make less contribution than if they'd done some things less well as it were. And so always to balance my efforts and I suppose that's why I've managed to do quite a lot in my life.

Also I'm quite lucky in that I have a short attention span! . So that whereas some people can only work if they aren't interrupted for several hours, I tend to get bored fairly quickly and perhaps work more efficiently if I turn to some other kind of work rather than switching off completely. I should say that some kinds of research, which I *haven't* done successfully, do require very long, uninterrupted time, for instance, preparing very elaborate computer code, something which I never done, and would be no good at. That *does* require uninterrupted time, but the kind of work I've done can be achieved with interruptions.

That means I've always worked in fits and starts and am more tolerant of interruptions from other people.That's made it possible for me to divide my time into small chunks more easily.

Look to the stars, and beyond

Astronomy is one of the oldest sciences - everyone has looked up at the stars since the dawn of history. But the history of the science is that we've gradually expanded our horizons.

In the old days, people knew about the planets. They pictured them something like this and gave each its (Latin) name - not all of them, as we now know, but even without telescopes pretty good - impressive actually on their relatative size and placing. But they thought the *stars* were in some sort of fixed vault of heaven, just a bit further away than the planets.

It was gradually realized that the stars were much further away than the planets because they were as bright as the sun, but appeared so faint because of their great distances.

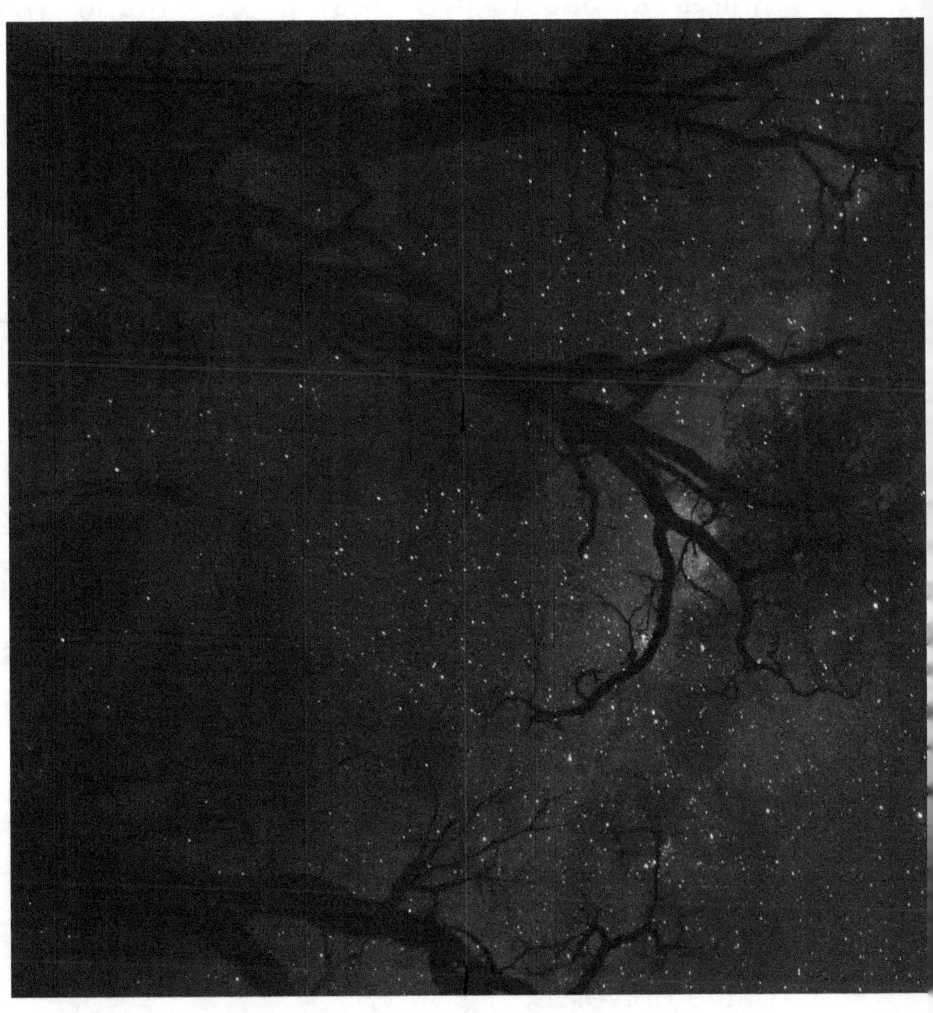

Then by the 19th century, it was clear that most stars were like the sun, and that the size of the universe was visible, that it was many, many thousands of light years.

The next big step forward was in the 1920s, when it was realized that our Milky Way galaxy was a huge disk of stars in which the sun was embedded. But it is just one of many. The Andromeda galaxy, which is about 2 million light years away, is like our Milky Way. It's a spinning disk of gas, viewed obliquely, with lots of stars in it, and it is something which contains about 100 billion stars. And in the 1920s and 30s, it was realized that these galaxies, fully equal to our Milky Way, are the basic large scale constituents of the universe.

But the trouble was, the telescopes at that stage weren't powerful enough to be able to see the

very distant ones.

And the history of astronomy really, ever since then, has been the history of more and more powerful instruments, enabling us to observe in greater detail, and, now, to greater distances.

The next big step forward was to realize that the universe is not static. It's expanding, distant galaxies are moving away from us. And the further away they are from us, the faster they are moving. This was originally discovered by Edwin Hubble, and it's been confirmed by much later observations. So we are in an expanding universe, where everything started off, crowded, much closer together.

And this raises two questions: what happened in the remote past, and what will happen in the remote future.

In the remote past, we would infer that everything was squeezed very close together. And galaxies couldn't exist in their present form. And so one of the aims of astronomy has been to understand the history of our universe in the past.

And here, we actually can do better than geologists can, because geologists can just *infer* (from fossils) how the past was, we can actually *see* it. We can see the past because since the galaxies are so far away the light is taking a long time to get to us. And so when we see a distant galaxy, we see it not as it is now, but as it was in the past. And by looking at galaxies, 1 billion 2 billion, 3 billion light years away, we can look back 1 billion 2 billion 3 billion years into the past.

And so we can learn something about what the universe was like earlier on, when it was more compressed. And what we infer is that galaxies formed when the universe was about a tenth of its present scale, and one   tenth of its present age.

We also have other evidence about the very early universe, about the time when everything in the universe was squeezed to a very hot, dense gas hotter than the center of a star.

We can learn something about this early universe in other ways too. And for the last 50 years, we've had a standard picture for the evolution of our universe from a hot, dense beginning to its present state. In this picture it started out very, very hot and dense, and expanded.

And after about 300,000 years, it cooled down to 3000 degrees.At that time, all the atoms became neutral, the electrons tied themselves to protons to make hydrogen atoms, and then the universe cooled down still further, and then galaxies started to form. So we have this so-called hot Big Bang picture of the universe.

Now, how far back can we go? If we extrapolate back when it was one second old, the temperatures about 10 billion degrees. And that's the time when nuclear reactions made hydrogen and helium, we can extrapolate back even further to when the universe was a nanosecond old. At that time, everything we now see with our telescopes, out to billions of light years would have been squeezed down to the size of our solar system.

There are many key questions about the universe. Why is it expanding the way it is? Why does it contain the observed mix of atoms and radiation, etc?

These can't be answered, unless we go further back still. And the problem there is that when we do go back to the first nanosecond, the conditions were so extreme, that we lose our foothold in experimental physics. For the first nanosecond, every particle in the universe was moving with more energy than can be produced with the biggest accelerators like the Large Hadron Collider at CERN. We don't know the physics.

But nonetheless, many people believe that many key features of the universe were imprinted when the observable universe was squeezed, not just the size of our solar system, but to the size of a tennis ball, yes that size.

This was when a process called *cosmic inflation* happened.

]

The physics is uncertain. But we do believe that we are seeing in the universe now relics of that very, very early stage. This is something people are working on. We don't know the physics, but we're trying to observe fossils, as it were, of that very early stage. We think that happened about 13.8 billion years ago.

What about the future? Is the universe going to go on expanding? And the obvious question then is, will it go on expanding forever? Or will it somehow stop expanding, and recollapse back to what some people call a Big Crunch? When everything falls together again?

That question was uncertain until two or three decades ago. But it's now pretty clear that the universe is going to go on expanding forever. We can't be sure, of course, because long range forecasts are never reliable. But the best bet on the basis of what we now understand is that the expansion will continue. And indeed, that the galaxies which we see with our telescopes, will accelerate away from us. So the Doppler shift, the redshift of the light, will actually increase.

So if we are observing a distant galaxy, and if we were able to live for billions of years and keep on observing it, we'd see this galaxy not only get further away, but accelerate away and then finally disappear from view. The long range forecast is that we will have a very cold and very empty universe, which will continue forever. And of course, all the stars in it will eventually die.

If we were astronomers observing in a 100 billion years time, or maybe 1000 billion years time, we'd see a very different universe. We'd find that the Andromeda galaxy and our galaxy would have crashed together, gravitationally bound to each other. And they would be the only things we could really see in the universe, because all the other galaxies would have receded from us, accelerated away from us, and they'd be as it were on the horizon.

So we would then be in a rather dull universe, very cold, very empty. Most , of the stars would have died out because the fuel in stars only lasts for a finite time. Only very dim, faint stars would be shining.

Even further into the future, all the stars would be dead, there will be no light in the universe at all. It would just be cold and empty. This would be a rather bleak future.

So there you are, that is the far future of our universe.

But then this raises another question. Is this cycle from the Big Bang to an ever expanding universe *all* the physical reality? Many people now don't think that it is. They have come to believe that our Big Bang was not the only one. And that *our* Big Band is part of some grand ensemble.

This is what the great Russian cosmologist Andrei Linde called *eternal inflation*. And this wonderful idea, which I think should be taken very seriously suggests that *our* big bang is just one of very many. And that will be others, which will lead to other cosmoses, which we can see there beyond the horizon. They will trace out an evolution, which could be like the evolution of our universe.

But more exciting still is that the physics in these other universes could be different from the physics in *our* universe. Some of them might have a stronger force of gravity, some might have quite different atoms in them. Some of them might be unpropitious for life, so they'd be sterile and still-born, as it were, some might have more interesting life in them than us!

So the story of expanding horizons, if we think of the history of astronomy, has gone through many phases. We started off knowing about our solar system and the vault of heaven. We then realized that our star, the sun, was one of billions in our galaxy, we then realized that our galaxy is one of billions that we can see with our large telescopes. And then, now, thanks to Andrei Linde and others, we are perhaps having a further Copernican revolution, realizing that the entire Big Bang that gave rise to the universe we can see is just one of many, perhaps an infinite number of big bangs.

So we now have a far grander concept of physical reality. This last step, the step from one universe to a multiverse is still speculative, but I'm hopeful. Just fifty years ago, we didn't know of the Big Bang at all, whereas now we can talk in detail about the very early stages of the Big Bang. So in fifty years from now, we will have the idea of a multiverse on a firm footing.

It may all be wrong, but if it's correct, we will be able to say something about physical reality, which will be far grander than what we can say today.

In these last few years, we have for the first time discovered a new force that is pushing the expanse of the universe and making it accelerate. As I say, we've learned a good deal more about the very first incipient stages of how galaxies and stars form the universe. And also discovered for the first time that the solar system is far from unique and that they are many other stars, orbit-al ng systems and planets, some of which may be like our solar system.

Now that's an entirely new subject, which is made possible by being able to detect very small motions in a star, manifestations of the pull exer-ted by planets orbiting around it.

So that's opened up an entirely new subject and 20 years from now, I think it'll be fascinating to look up at the stars that we are familiar with and realize that each one of them has planets orbiting it and be able to see what the properties of the planets are just as we tell children now about the planets of our own solar system.

Our amazing world

We cannot fully understand the origins of that huge cosmic complex of the stars together with the minute DNA double helix molecule that en-codes our genetic heritage. We also depend on the sun and its power. And then there is so much more beyond the Solar System that we can't un-derstand without taking account of that ultimate cosmic complex with which we are intimately re-lated. And then too there is the quantum of the quantum world of the atoms and subatomic particles with remarkable interconnections between the micro scales and the macro [???] world..

Our everyday world is determined by atoms, their combinations of molecules, minerals, living cells and the starshine that depends on all of that. Galaxies are probably held together by the gravity of a huge swarm of subnuclear particles.

The tremendous time spans of evolution give another perspective on the question of why - and how - our universe is so huge.within all of that our world is so small.

It's taken four and a half billion years for the emergence of human life on earth. Even before the sun and its planets were formed, earlier stars must have transmitted oxygen transmuted into carbon and the other atoms over about ten billion years.

The size of the universe is about the distance travelled by life since the Big Bang. And so the present visible universe must be about ten billion light-years across.

Is that not amazing?

You can see how unimportant we are in the cosmic scheme. We're a consequence of a long chain of events that extends back even before our solar system was formed, long long before our arrival on the scene.

We may want a deeper explanation of how the universe is as it is.
Einstein once said the most incomprehensible thing about the universe is that it is comprehensible. We looked at the laws of physics that apply not just to the earth and the solar system but to the remotest of galaxies.

Now, you know about gravity through Newton. It's what makes apples fall and how's the moon and what holds the planets together in their courses. That same force holds the delicate Andromeda galaxy and started the black holes that   will make it collapse.even the most distant galaxies are just the same as those in that we study. It is possible indeed that what we kniw as our uni- verse is really part of a vast multi- universe. In fact all the parts of the universe are evolving in a similar way as if from a common origin.

Looking at this uniformity - that is where we have cosmology, always of course coming on new mysteries..

I think we're at an exciting time now, not at all the plateau some people suggest. Subnuclear phys- ics is a subject that depends on experiments that are very difficult to do, and has been held up for 20 years through the lack of new crucial experiments. With the new accelerator CERN in Geneva comes the hope that we'll discover a new set of particles, predicted but not yet discovered. But it's true that in that subject, things have stagnated rather, because with a lot of ideas, there's no way of testing them. That's the downside of positive achievement because its great achievement is the idea that particle physics can understand everything that is observed - except things that need billion dollar experiments to test!

Doesn't that suggest that the subject has developed a certain maturity?

There is an important point to make here, though, that this is only one branch of science, even only one branch of physics. Understanding the very smallest basic ingredients of the matter is clearly important and fundamental. But it's very misleading to be a total reductionist.

People sometimes say, well, it's like a building. You've got particles at the bottom, then the rest of physics and chemistry and cell biology, then organisms and then social science, and the economists up in the penthouse. You can think of it as a sort of hierarchy.

On the other hand the analogy of a building is not a very good one, because in the case of a building, insecure foundations can wreck the structure, but that's not the case in the hierarchy of the sciences. Biologists aren't impeded in the slightest if we don't understand what goes on inside an atomic nucleus. They *are* impeded by the complexity of what they're trying to understand.

So each science, each level in the hierarchy, has its own irreducible concepts and tries to interpret things in terms of those concepts. You've got to solve the problems that beset science in terms of *that* science's concepts. A very important principle.

And that therefore means that although understanding the subnuclear world is important as a branch of science, it's not something which underpins the rest of science directly. So, the rest of science is not being held up through not understanding some bits of nuclear physics.

There *is* one science which *is* held up, however, and that is understanding the very beginning of our universe. This is something I work on: how the universe evolved from simple beginnings to its present state. And we can extrapolate back with a good deal of confidence to when the universe was about a billionth of a second old.

That must seem a pretty bold statement. But if you measure on a log scale, a lot can happen in the first billionth of a second. And the reason that we can't understand this tiny, tiny incident is that if our extrapolation backwards is correct, then conditions get more and more extreme as we go back to hotter and denser states. The laws of physics that we are confident about break down or get transcended when you get back to this early stage because the particles are all moving around with high energies and reproducing with huge acceleration.

So for the very beginning, we need somebody who knows fundamental physics. And there is this interesting link between the very large and the very small because in most of science, we are not impeded by the fact that there's no unification between the quantum world, which is the atom, of the microscope, and the cosmic world where gravity dominates.

The reason for that is that in the case of single atoms and molecules, gravity is not an important force. Gravity is not really important when you pack together huge numbers of particles. And so, conversely, in the case of planets and stars, where gravity *is* a dominant force, quantum uncertainty is an important concept. So this is how we are: the quantum fuzziness in the orbit of a star is negative because apparently so big compared to an atom. So you see science is not held up by the fact that we have used two disjoint ideas in modern physics, the micro level of the quantum and the cosmos, and cosmology where Einstein's theory hold sway.

But right back at the very beginning, when the whole universe was squeezed within the size of an atom, quantum fluctuation could shake the entire universe. So, clearly, you *do* need to say what unifies the very large and the very small to understand what's going on. And just till we have that theory, we won't be able to answer these embarrassing questions like what happened *before* the beginning.

We suspect that the answer to those questions is that the whole idea of space and time has to be revised and extra dimensions brought into play. But right back at the beginning, we in cosmology, who normally think about very large things, do need to take cognizance of the very tiniest structures and therefore we are very interested in developments made by people studying that.

And then we wonder about other domains where light had time to reach us in the ten billion years since the Big Bang happened. Not just millions of times further that what we can see now but millions of powers of ten more.

And yet more still...

Maybe the laws of geometry are different in other universes. It is so complex, so connected.

But - does it have a future?

So here we are on Spaceship Earth, hurtling through the void with its passengers fractious and anxious.

There are indeed long term issues here that we need to address. But we are not addressing them, partly because of minor disputes which we can't settle. Sothere's a big gap between the way things *are* under current polices,   the way things *should* be.

What are our biggest dangers?

I'd say *not* an asteroid colliding with the mother ship - us. It's true that such risks exist, but they are getting no bigger. They were bad for the dinosaurs, they are bad for us, but they are most unlikely to happen in the next century.

What I worry about most are the kinds of risks induced by us humans, either collectively on the climate or the environment, or individually because technology is so powerful that even a few people can have a catastropic effect that cascades globally.

I look at it in astronomic perspective. The earth, remember, has been around for 45 million centuries. This is the first century when one species, namely the human species, has the future of the planet in its hands. What we do in this century will determine whether we leave a depleted planet for future generations or whether we can survive sustainably.

To repeat, because it matters. I am an astronomer, I think about long times. And I know that the earth has existed for 45 million centuries. But this century is special.

And it's the first century when we've understood our place in universe, the first when people have left the earth.

And it's the first when we have the power to determine the future. We saw how people landed on the moon 45 years ago, I'm old enough to remember that. And I thought at that time, that by now, by today, there would have been many footprints on Mars. But, as I mentioned earlier, there hasn't really been the motivation to send people rather than robots.

I think that'll change. I think some people now living, maybe some of you reading this, will walk on Mars as an adventure - an *adventure*, a step towards the stars. But when you go to Mars, don't expect it to be comfortable! it's worse than being at the South Pole or the top of Everest, because this earth is a very special place. There is no future for mass emigration from the earth,, we've got to solve the world's problems w*here we* are - *he*re.

And if you do go to Mars, look back on the earth, and remember just how special it is. You'll want to preserve it even more. So I hope some people reading this may walk on Mars. But if you do, then you may still think that this pale blue dot in space is very special.

So let's all look after it.

I try to be hopeful, of course. But I think we are going to confront a whole lot of new challenges because this is the first century when human beings collectively and singly are going to have an effect on the entire planet.

Collectively, we are well aware of the consequences that will have to - you know - change the climate, destroy the Earth's diversity and so on, . It's good that this has come up the political agenda as a long term concern.

An issue I worry about is whether we can cope with the rising population and the aspirations of the developing world etc without leading to reversible long term deleterious effects on climate and on space generally. So that's one issue.

The other issue is that we are confronted by a new set of risks from the fact that we are living in a more and more interconnected world where individuals are more empowered than ever before. This means that society is now far more vulnerable. We've always accepted that there are risks, and that a new technology is risky, but we now confront a new class of risks, which are potentially more catastrophic.

The risks from just the use of new technology could be catastrophic. At one time in their earlier development lots of boilers exploded, and that was horrible for those nearby. But there was nothing *global,* It was just horrible locally. But now we can imagine technologies, biotech of some kind, leading to some hazard felt *globally.* And even though the probability may be very low our concern is affected by that probability multiplied by the consequences, so if the consequences are really catastrophic, we should still be worried.

So that's a new class of risk we need to worry about - some sort of unintended consequence of new technologies in, you know, the interconnected world.

And also, of course, there's the risk of terror of various kinds. I'm concerned about the great empowerment of individuals who will be able to use terror not just as an additional way of causing hazard or habit, but using new technology. I'm thinking not just of people with the mindset of those who design computer viruses, but of those who may actually design real viruses by error, a new cancer threat.

The question is how we can cope with these and what hopes we can have that these technologies will be applied to human *benefits and with great caution.* The 'Global Village' is likely to have its village idiots and we will have to cope with them, because even *one* could be one too many.

This I regard is a very practical problem - dealing with the empowerment of individuals. I'm not just thinking of the traditional terrorist that we are concerned about now, the fundamentalists, etc. they can be all kinds. I mean, you could imagine some eco freaks thinking that the world would be better off off without human beings or some if the the strange science-based cults.

And so I think there's actually a real new risk from people who are confident in science, but lack a touch of reality, or are obsessive or extremist and such like.. And so that's something we may have to cope with.

So, in summary, I think we have the collective consequences of our actions, which can be catastrophic for the biosphere, and also the consequences of individual error and terror amplified in a world that is more interconnected, and where individuals are more empowered.

So for those reasons, I think the 21st century will be an especially hazardous one, where we will have to confrontv not just additional risks, but a new set of risks. whats more technology is advancing faster than ever and on a broader front, one you couldn't have predicted.

And if we couldn't have foretold what we now know about the world today, still less can we predict the world in 100 years from now on the basis of what we can do today.Technolgy is advancing faster and faster. But, on a broader front, I would say that we're changing the world in new ways which human beings haven't done before.

One thing that *hasn't* changed over the thousands of years of human history has been basic human physique and human nature. But of course, with new kinds of drugs, even that may change and for the very first time. That too is a new kind of uncertainty, a kind of dangerous feedback into society.

And so for all these reasons, I think we are going to have challenges which politicians will find hard to cope with.

Does being an astronomer have any special impact on my views? My answer would be essentially no. I think I'm just expressing the views of someone who is a scientist and in touch with scientists and has a feel of how science is going. But I think there is one distinctive perspective which scientists bring to these issues, if they are familiar with cosmology and astronomy. And that is, they have an awareness of the long term future.

Yes, it's up to us.We *could* cause a catastrophic extinction. But let's hope we don't.

Getting into space

What I've been describing is a good example of how the cosmos has become much more interesting as a body of data. Our understanding greatly enriched by new technologies.

And there is more!

Think about space in the widest sense. Space technology has developed enormously, driven by the pressure of events and the requirements of rockets and so on,

These developments have been crucial for my branch of science. We've understood a lot more about the solar system through actually sending things there. Also, by sending telescopes up into orbit, you can observe in a way that you can't from down on the ground.

There are two advantages of going into orbit. One is that by getting above the blurring effect of the Earth's atmosphere, you get much sharper images. And that's why the Hubble Space tele-scope is so important. The other is that there are certain kinds of radiation, ultraviolet, infrared and x rays, which are emitted by cosmic objects. They don't get down to ground level because they're absorbed in the atmosphere. By going up into space, you can observe in different sort of windows, as it were. If we go into space, we can see a wider range.

Being on the ground is like being able to detect notes in a certain fairly narrow range of pitches. But we can learn a lot more - detect more - by going into space. So through space technology, we have been able to extend our knowledge and a lot of discoveries to be made. The ones I worked on about extreme cosmic explosions would not have been possible had it not been for data gained from equipment in the space above the Earth's atmosphere.

So that's the important part of my subject. But of course, that's only been made possible as a byproduct of the development of the technology, which is driven by other imperatives. Obviously, space technology developed in the Soviet Union and in the US because of superpower rivalry. And of course, it has continued to be developed for military purposes, but now also for commercial purposes. Now for everyday life we depend on space - for telecommunications, GPS, position sensitive detectors, and all kinds of other things.

So space is very important, technologically, and I think that by now civilian expenditure on space actually exceeds the military. And of course, scientific expenditure is a tiny spin off of all that, but nonetheless very important.

Looking ahead to the future of space, there are plans being made for large instruments which will be able to study very distant objects and therefore see back in time and see back to when the first thousand galaxies were forming. That's because the further away you look the further backward you're seeing in time.

These new instruments will be able to detect planets around other stars, even planets like the earth. That'd be very exciting because the planets we have detected so far are big ones rather like Jupiter and Saturn, the giants of our solar system. We would really like to see evidence for *any* kind of planet, including some more like the earth. Within 20 years, we will certainly have telescopes in space that will be able to image a planet like the earth orbiting another star. It won't be a really sharp image, but we'll be able to learn quite a lot from it .

To illustrate what we could learn. Let's imagine you are an alien looking at our solar system from say 30 light years away. If you are looking from that distance the sun would look very like a star and the earth would be about a billion times fainter. It would look, in Carl Sagan's nice phrase, like a pale blue dot orbiting around a star very close to it in the sky.

If you were an alien and had a very powerful telescope you might be able to pick out this earth which would be rather like a firefly next to a search light because it would look so close in the sky to its star, our sun. But if the aliens could look at this pale blue dot, even if they could just see it faintly, they could learn quite a bit about it. They could learn about it because the shade of blue would be slightly different, depending on whether it was the Pacific Ocean or the Eurasian land-mass facing you. And if you kept on watching you could learn the length of the year, of the day, and the seasons and so on.

Well, 20 years from now, we'll be doing that sort of thing for stars like the sun, to see if they have planets orbiting like our Earth.

So we know @something about the topography, deep in the atmosphere of those planets. The question then will be, *life* in any of those other worlds? And that's a separate question. It's a question for biology - a much harder subject - and astronomy. The question of the origin of life isn't really understood here on earth. But until we understand *that*, we won't have a feel for how life got started. And *that* question, of course, is one of the great challenges for 21st century physics.

So if we look at what space science is going to do for us in the next 20 years, one thing is going to be allowing us to study other earth-like planets and stars. Now look at the development of space, science and technology - it has been mo-tivated by not only sending forms into space and restricted space. It's also been motivated by sending *people* into space.

And of course, the high point, perhaps the greatest space spectacular ever, was the incredible American moon landing (heard of that?).

And that, of course, was a crash program, driven by superpower rivalry, where the Americans made huge efforts to send men to the moon.

I am old enough to remember that exciting episode.

But of course, it's interesting to note that the last man on the moon was there many years ago by now . And so, to those people who are young

today, men on the moon is past history. That's syndicated to school kids now, they know the Americans landed on the moon, and the Egyptians built the pyramids because both seem really arcane national goals which by now may seem hard to understand the motives for.

But those moon landings were not just a surface thought at the time, the step towards a program exploring the planet by sending people to Mars, etc. It was a one-off episode, motivated by superpower rivalry; but then there was no motive to continue with the immense resources needed to continue it.

That does raise the question of whether people will return to the moon in what is a long term future for people in space.

There are lots of debate about this. Some people say that it's very exciting to send people into space. And if we look at what's happened in the last 10 or 20 years that's what's made the headlines. There's been lots of publicity for that. We have the pictures sent back from the surface of Mars, from the moons of Jupiter, from space probes. But the *human* expansion of space has only made the headlines where there's been a disaster when the shuttle twice crashed. And that's caused public interest - naturally.

But the International Space Station, which is circling around around the Earth, doesn't really inspire the public very much, because more than 30 years after people walked on the moon it's not very exciting that something should just be going round the Earth at huge expense.

And so the question really is whether there is a future for manned spaceflight at all. The Americans have enunciated that their long term program is to go back to the moon and go to Mars, but I'm unsure whether that will ever materialize. It's *very* long term.

The American program is also very risk averse. The shuttle accidents have been a sort of national trauma as it were, even though the shuttle has a 98% safety record. And I think one of the problems with the American NASA Space Program is that they have presented it as though it's a low risk venture. And therefore, it has been followed by delays and inquiries and extra expenditure, etc. I mean, it would be far better if they accepted that space can be very dangerous in regard to these people that are like intrepid test pilots, and accepted there'll be accidents. And I think that's going to be hard for the American public, for politicians, to accept.

So my personal view is that the only future of manned spaceflight lies in a different approach - a higher risk, private enterprise approach when costs come down. And when costs are cut private enterprise can finance the venture. And people will accept much higher risks that ways than the American taxpayers are prepared to impose on the public to finance astronauts. And people will have to be willing to go to Mars with a one-way ticket.

I think that if it was funded by private enterprise, the same as Formula One racing is, or by individual people, and the people who went on these expeditions were regarded as high risk adventurers like mountaineers, balloonists and such like, then there may be a future for manned space-flight as an adventure.

I think in other words that we've got to emphasize it as *adventure.* Because the practical and scientific case for sending *people* into space gets weaker and weaker all the time with each advance in miniaturization and robotics. We can now produce more and more sophisticated robots that can miniaturize things and so many of the activities which once needed people can now be done without. And the cost is far less, obviously, if we don't send people.

The European effort in space has been rather low level compared to the United States. The European Space Agency has a budget about a third or a quarter of theirs. But my personal view, as a European, is, as you'll have gathered, that we in

Europe should eschew manned spaceflight completely and spend all our budget on the advanced robotics developments in science. The Americans spend three quarters of their larger budget on manned programs. If we decide we're not going to do that I think we can do very well, just as the leading Particle Physics Lab in the world is now in CERN Geneva. So we can make the leading space science an activity that is driven from Europe.

I'm reassured by how science is taken 'pretty seriously' in the UK. We have a good record of supporting science. Also, we link it to policy – for example, in the regulation of embryo research – which we do better than in mainland Europe or the USA, as parliamentarians have engaged with scientists.

In more enteral terms I'm quite reassured by how science is taken 'pretty seriously' in the UK. We have a good record of supporting science. Also, we link it to policy – for example, in the regulation of embryo research – which we do better than in mainland Europe or the USA, as parliamentarians have engaged with scientists.

So many things are good, and it's a good time for young people to get into science.

Life elsewhere?

Unless you are a creationist, you're happy with the idea that we are the outcome of 4 billion years of evolution. By now we can trace back 13 or 14 billion years to the Big Bang when the first things were made.But even people familiar with that tend somehow to feel that we humans are in some sense the culmination, the end, of all.

In fact as an astronomer I know that the sun is less than halfway through its life. It will flare up and die in about 6 billion years from now. So yes, it's less than halfway through his life and any creatures that witness the demise of the sun won't be humans, they'd be different from us as we are from bacteria. That therefore means that if we have unintended consequences in the 21st century, which destroys the future of life on Earth, then those consequences will resonate far far into the future, to even what you might call a post human era. So in a sense, if you are aware of the long term future of life here on earth and life far beyond, then you have an extra reason for concern over and above what concerns us, which is our immediate descendants. So that's the only extra perspective which I bring.

The cosmic importance of what happens here on Earth depends on another fascinating question, which is the kind of thing taxi drivers always ask me if they know I am an astronomer, Is there a life out there? That's another question I'm often asked.

So - *are* we alone in the universe?

Well, I'm prepared to believe that we are not, In fact I've been involved in a project searching for life on other planets by looking for evidence of machines there, as only other life could have created them. He reckons there is a 1%–2% chance of finding machines, although the chances of there being life elsewhere are higher. 'We don't understand the origin of life yet, so we can't tell whether it was a fluke or something that was inevitable and so would have happened elsewhere.

And the answer to that question is we just don't know. It's a biological question. And we don't know how likely it is that life got started. We don't know how likely it is that evolved into something called intelligence. You can see what other kinds of life might exist. But, of course, if life is already widespread, then we are just a tiny fragment.

Whereas, if life is a very rare phenomenon it could be that our tiny experience is cosmically important because it may be the one place where life got to this state.

These questions about the origin of life and of life in other places are among the most exciting of subjects, bringing together all the disciplines, because you need to have experts in astronomy and in physics, and biology, in order to address those kinds of questions.

.

The responsibility of science

To me, two important goals nowadays are the education of scientists and - a separate goal - the education of the general public.

I think we all realize that the future of our society depends on technical innovation that could be better directed than it has been up till now. And therefore we do need people who are going to be able to understand science at the highest level and engage in research, medicine and environmental science, etc. And certainly from a nationalistic perspective, we in the UK will not have a bright future unless we can maintain expertise in science and technology in the face of the growing competition, not just from the United States, but from the Far East. So yes indeed it's very important that we in the UK maintain standards of education and research for specialists in science and technology

A quite separate issue is of course, the wider public. We're not just dealing with professional scientists in the specialist field of science and technology but the public in general.. And this is also important because young people - you maybe - are going to grow up in a world ever more moulded than ever by science and techno-logy. We - you - are going to be faced with more and more ethical choices about how to apply sci-ence and, obviously, applications of genetics and biology and how we get in with the con-sequences of our collective actions on the atmo-sphere and global warming and diversity etc.

These are issues which are tactical, which in-volve ethical and practical judgments. We need a public that is better informed to be able to participate in debates on these questions because it shouldn't be just the professional scientists who decide on the applications of science. *They* have no special ethical sensitivities.

What is important is that scientists should feel an obligation to explain the scientific background to the crucial decisions society and politicians have to make, and participate in dialogue with the wider public. But the wider public too obviously has to have some feel for science, if that debate is to get above the level of tabloid slogans. And so I think it's very important that everyone should have *some* sort of feel for science, in order that we can make democratic decisions on how science should be applied throughout this century.

And, obviously, it's the key ideas about the details that we can hope that people will understand. And also the fact that we are never really certain about things.

Scientists often find it hard to explain. We can be pretty sure, though, that there is nothing that is ever risk free. The worst kind of questions reporters are being asked is, 'Is there a risk from x or not?' Say yes or no, you can never answer that question for sure. Whether it's the MMR vaccine or anything else, you've just got to say that the risks are very low or whatever and that we should listen to this batch of experts rather than that batch of experts, etc. So I think it's essential to ensure that the public has a feel for what science can do, what the issues are, how they depend on scientific knowledge, and what the uncertainties are. If so there can be a real dialogue extending far beyond the scientific professions about how we ensure that the scientific discoveries that are being made are for the broadest national and international benefit.

So I think, both when I was President of the Royal Siciety and now, that all of us who are active scientists, and academies and universities too, need to be concerned not just about maintaining the standards of education for specialists in the subject, but also in engaging in outreach to ensure that the public is aware of the scientific issues.

In other words I think it's the job of scientists to do all they can to get the word out to the wide public, especially on issues like climate and health where it matters and where the public has to decide. On the political front I haven't myself been too active in party politics recently. In the House of Lords I'm a   Cross-bencher, that is, someone without party affiliation. people who are supposed to bring some expertise and commitment to the place but you don't participate as a member of a political party. I would rather like to be involved more because unlike most scientists, I've always been fascinated by politics and think political decisions matter and that scientists should indeed take part in them.. And so I would like to become more involved at the moment, but because of other commitments. I haven't been very active and don't feel I pull my weight adequately. I try and attend the Lords twice a week, but that may be just a brief attendance

I would hope that in due course, I can become more involved. What's frustrating is that there are often debates or activities I'd like to participate in but they're changed at the short notice and because you're doing other things, you can't be there. So I can't participate in the way that the circle working peers do that. I hope that in a few years, I will be able to participate more.It's a great privilege to be there.

Mind you I think scientists sometimes bemoan public ignorance of science too much. It's just as worrying that people are ignorant of history. I mean, you can't have a well informed opinion on what's happening in the Middle East without a bit of background or on what's happening in the Balkans, etc, without having some knowledge of history. It's equally sad if, say, people don't know the history of their country, if they can't find South Korea or Iraq on the map. Many people in the US and the UK can't do even that. And yes it's also sad if someone doesn't understand the basics of science because then they wouldn't be able to have a proper informed debate about the kinds of issues which have a scientific dimension but go beyond science and requires ethics and politics and economics.

So I think we should equally bemoan the fact that the average person is uninformed about the historical background to those key arenas, as we should bemoan the fact that they may not understand the basic science.

I think the new communication media or the internet are crucially important in public education.I mean if a scientist really wants to get their ideas over, clearly writing a book or an article is the surest way for this, and then intermediaries get your words out there. I've had some experience myself of television for one of the main terrestrial

channel channels and found it frustrating because of the compromises you have to make -
the average viewer is surfing the channels and
can't be expected to pay attention for a whole
hour. And that leads to a lot of repetition, etc. and
a lot of distractions. So it's very hard to present
any serious ideas on the main TV channels. But I
do think that with the internet it *is* going to be
possible to make videos available to to convey
ideas to a large audience of viewers. And this
may be helped by the possible revival of serious
documentaries.

Also *dialogue* is very important. I think scientists
have often been rather remiss in being unwilling
to get into the arena and get involved in dialogue
with the public, and with the others who participate in intellectual life. We don't really engage
with them as much as we should.

I think that's worrying for two reasons.

First, what scientists have to say is fascinating to a wider audience than just the specialists. But it's important for the image of the scientific profession, that its practitioners shouldn't be seen as sort of standoffish or elitist, they should be willing to participate in discussions and get involved in politics and so on. You could say that that's the essence ofwhat's meant by the public responsibility of scientists

I'd say that the classic case of this might be the atomic scientists of World War Two, who as we know now in great secrecy developed the atomic bomb. Many of the physicists who came from the academic world and worked on the bomb project during World War Two, went back to academia afterwards. But they did maintain a long term concern, feeling a responsibility to do all they could to urge steps towards arms control. They were involved in high stakes indeed and I think many of those people did in fact take a very responsible action.

I've been privileged to know, some of the people who were involved in that Los Alamos work ; most of them are sadly, no longer with us. There's no realm counterpart of that in the younger generation. But people like Joseph Rotblat and Hans Bethe who were involved in Los Alamos, devoted the rest of their lives, or anyway a large part of their lives, to campaigning to control the powers they had helped to unleash. And I think they set examples. They they knew very well that the decisions on nuclear weapons were political decisions but they felt that they had a special obligation as scientists.

To give another analogy, you may not be able to *control* your teenage children, but you're a pretty poor parent if you don't care what happens to them. Likewise, if you're a scientist and your ideas are, in a sense, your intellectual projects, you should care what *happens* to your ideas, that they should be used for human benefit, and that the risks of their misuse should be minimized. So scientists do have a responsibility.

Then there are the Pugwash type activities that I've been a bit involved in. Theyre called PUG-WASH because in 1955, I think it was, Bertrand Russell and Albert Einstein signed a famous Manifesto drafted for them by Joseph Rotblat. And as a consequence of this, an institution was formed to get together scientists from East and West and Joseph Rotblat who instigated this movement remained it with it for fifty years until his death. The first meeting of this organization was funded by a Canadian millionaire call Cyrus Ethan and took place at his home village called Pugwash in Nova Scotia in 1957. And the conferences therefore, culled the name Pugwash conferences' and two or three hundred have been held since then.

The Pugwash Congress was especially important in 1960s because at that time there was very little opportunity for interaction between scientists in the Soviet Union and in the West. But they were able to meet in an informal way at these Pugwash conferences, and then they reported backto their governments. I think the record suggests that in the 1960s, these meetings were helpful in lubricating the steps towards the partial Test Ban Treaty in 1963 and in leading to the Anti-Ballistic Missile Treaty in around 1970. Subsequently more channels opened up and the agenda became wider than just the superpowers confrontation.

So Joseph Rotblat had the satisfaction of seeing that in his last years his goal came to seem less utopian. Sontoo forb Rob McNamara, who, in his later years, produced a film called *The Fog of War*. He was convinced that we came very close to catastrophe at the time of the Cuba crisis and that the only long term hope was if we could rid the world of nuclear weapons. So it was not just people who could be dismissed as idealists like Joseph Rotblat ,but also people who were more hard nosed, in their background anyway, like McNamara, who came to believe that we should strive harder towards this goal and who were concerned about nuclear proliferation and that nuclear weapons were in a special category. In order to discourage proliferation, those powers had an obligation to cut down their own stock-piles,

Thoughts...

I think all my work has involved trying to use the physics that we understand to make sense of what's out there, and how everything evolved from some stage beginning into the cosmos that we see around us and of which we are apart? And I suppose, compared to most people in my subject, which now tends to be subdivided into specialities, I've worked on a very broad front and tried to remain involved in several areas. So there are two or three areas that I'm still active in and that I go to conferences on.

One of them is trying to understand how the universe has changed from being a sort of mix of amorphous gas, which it was, in the first million years after the Big Bang, to its present state when it's very diffuse, full of stars, galaxies, etc. and trying to understand how the first galaxies and stars formed, what they were like, how we can use observations to learn something about this and to test our theories and to see which of those competing theories is closest to the truth. So to understand cosmic evolution from its amorphous beginning to its present structures, why galaxies exist, why they the way they are, etc. - that's one thing.

The second thing of my work has been really to understand extreme phenomena in the universe. Because another reason why we do astronomy is because *nature* has performed experiments we can never perform in the lab, extremes of gravity, extremes of temperature and energy, etc. And so, by looking at things happening in space, you can perhaps learn something about the laws of nature under extreme conditions, which you could never simulate here on earth. And that lends added interest to this phenomena. And I've done quite a lot of work in the last 10 years on some of the most extreme phenomena, where there's releasing in a fraction of a second more power than the sun puts out in your lifetime - a process which involves huge energies, particles moving nearly at the speed of light, and probably the formation of black hole and other exotica.

I'm trying to make sense of this. As I say we haven't got very far yet with this, but I find the process fascinating and perhaps get clues to the nature of the natural condition through the theories of black holes, etc. So that's one set of ideas I'm thinking about now.

So, the two main areas are, how the structures were that originated the universe, and how we can learn from these extreme phenomena in the present day universe more about the laws of nature and about gravity.

I also do some more speculative things about the nature of the laws of nature themselves and wonder if there is part of universe where we can observe *the* physical reality or just some tiny fragment of it, perhaps the big bangs, etc.

There are the more speculative things to work on too. I love what Stephen Hawking said - something like that it is not implausible to believe that we may be on the edge of understanding the underlying and unifying laws of nature and of the universe. That's more   hope than   statement you'll say.

Well yes, it *is* hopeful. I think it's good that some people are hopeful because that motivates them to try harder- essential. But I think there's also the far-ranging question of whether we will, in the next few decades, complete the program we started with Newton and continued with Maxwell and Einstein, of trying to unify together the different forces of nature. Electricity, magnetism, nuclear forces, gravity, etc. Some people are hopeful, some less so. And so this search fir the first fundamental theory of unified theory is an important goal of science.

Two things about that,

First, it could be that we can never have this theory because it's beyond human brains. After all there's no particular reason to believe that human brains are matched to the understanding of the deepest level of physical reality, just like my dog can't understand quantum theory. So we've got to accept that there may be a theory that is beyond us humans. So that's one thing.

The second point, which is often misunderstood, is that if we had such a theory, it would be the end of a certain style of science. But it would not in any sense be the end of science itself. None of the scientists I've mentioned are held up in the slightest through not understanding what happens deep inside an atomic nucleus - that's irrelevant. They're trying to understand what happens when atoms are packed together in the big concatenations of organisms. That's the complication for them.

And so, the unified theory, if it exists, if we find it, will be an important step forward in the program goes back to Newton: to unify the laws of nature.The force that holds the moon in its orbit is the same one that makes the apple fall and touch the ground. But although this will also unify the physics of a small quantum theory, the piece of a large gap in Einstein study, the third frontier in science, the very complicated and the most complicated things are neither as small as atoms nor as larger stars, they are complex things like us.

The greatest challenge to science is to understand that very complex things ng.

Questions about God (whatever word you want to use)? People often ask me about all that too.

Well actually   my views are very modest and therefore very dull. In the sense that if science teaches me anything relevant about religion, it teaches me that even something as simple as an atom is pretty hard for most of us to understand. And that makes me very suspicious of anyone who claims to have anything more than a very incomplete, a metaphorical, view, of the very deep aspect of reality. I am skeptical about claims to reveal dogmatic truths about things of massive importance, and so I can never be a firm adherent of any dogmatic over-arching belief. Nonetheless, I believe in peaceful coexistence between mainstream religion and science. I myself have grown up in the tradition of the Church of England. So that's   part of my culture, part of my country's culture. It has tremendous aesthetic resonance for me, and social resonance. And so I'm very happy to participate in the rituals of the

Church of England. And so I do attend, I call it chapel as it were. And I do this because I believe it's important, I would be sad to see it decline. But I fully accept that if I had been born not in this country, but, say, in Iran, I'd be going to the mosque in the same spirit. So I don't believe in any sort of absolutism.   I accept that religion is culturally conditioned, but nonetheless accepted as important.

When I was Master of Trinity College Cambridge, weekly attendance at college chapel was a 'happy' duty. I would have loved to have been a composer – I wish I had the musical talent – yet life as a Cambridge astronomer has its musical compensations: amount of music performed routinely every day in Cambridge is hugely impressive and important. I've never been very religious, so it isn't science that has turned me to atheism, just that, as I say, I knows how hard it is, as a scientist, to understand and explain even relatively simple things I am reluctant to accept dogma that claims to have found all the answers.

I definitely feel allergic to fundamentalism, which I believe can be counterproductive for two reasons.

First, I believe that fundamentalism is a threat to all of us, like the fundamentalism of the deep south. And Islamic fundamentalism is also the fundamentalism of a crazy new age sect. They may talk the language of science but many are cut off from rationality and could be dangerous. I think we need all the allies we can get, from whatever source, against this.

I also think that we should emphasize that it's possible to practice a religion and be a scientist as well.

Otherwise we wouldn't have science because if you were, say a six former in a British school and devout Muslim, and someone cane and told you that you can't believe in Darwin and in your God, then you're going to choose your God and we'd have lost the science. That would be bad for science, bad for all of us, and so I think to preach the possibility of coexistence between science and religion is something I wish to do. It's for the sake of science too because to hold the opposite means that people who would otherwise like to participate in science in the context of a world culture will be detered from doing so.

Over to you

My advice to a young graduate starting out in science now?

I'd say that your risk as a young scientist is to look at all the journals on the shelves and be tempted to think that all the big problems been solved. And so the important point to make is that if you are going to make a cintributiin you're not going to be - or shouldn't be - pursuing the old problems that earlier people got stuck on. You'll be addressing questions that couldn't even have been posed 20 years ago.

So you want to choose a subject where things are advancing fast, new problems that couldn't have been posed 10 years ago. And also where there are new techniques, more hack in the computers, more powerful instruments, etc. So you can have access to data that your predecessor didn't. And then you can make a contribution that gets you to do something which is new and which the older generation never had a chance to do. That's one bit of advice.

But the other bit of advice I is, if you really want to do something, don't be deterred by negative advice from old professors like myself!!

Acknowledgements

The material in this volume has been selected and adapted with the author's permission from earlier versions (orincipally interviews and lectures) such as https://sms.cam.ac.uk/collection/2200350/, https://www.ted.com/talks/martin_rees_asks_is_this_our_final_century/ *https://www.iop.org/resources/videos/lectures/page_59756.html#gref/* https://www.youtube.com/watch?v=NpHz0jkJ-zoo/

More things you might like

If you enjoyed this book you might also like:

A History of The Royal Society, https://royalsociety.org/about-us/history/

Walter Isaacson, *Einstein: His Life and Universe,* Simon and Schuster, 2017.

Martin Rees, *From Here to Infinity: Scientific Horizons,* Profile Books, 2011.

Martin Rees, *Just* Six Numbers! W&N, 2015.

Martin Rees, *On the Future: Prospects for Humanity,* Princeton University Press, 2018.

Questions for discussion, debate or action

When you're being creative (in whatever sphere) how do,you and/or your friend's work? Any re-semblance(s) to the account here?

Would it matter to you if the universe/multiverse did end up cold, and dead and lonely? Why or why not?

What do you personally think of the view that science and religion are incompatible? Or, to put it slightly differently, that you cannot be at once a proper scurntdt *and* a cimmutted member of a re-ligion? What do you yourself think are the argu-ments for and against these views and have you heard them from others?

It is clear that Martin Rees was at first unclear what career he would follow?does/did this apply in any way to,you, and if so what are/were your possible options and why?

"What scientists have to say is fascinating to a wider audience than just the specialists". On the basIs of this book would you agree? *Were* there things in it that you found 'fascinating'? If so, what? and would you be able to pass them in, in some form at least, to,others?

If you yourself are thinking (or did think) of going into some branch of science do you find Martin Rees' advice here helpful   to you personally? Why or why not?

"More and more powerful instruments ... ". What do you think might be coming next, and might *you* be a part of developing it?

Would you recommend this book to your friends or acquaintances? Or to your (or any) institution's library? Why or why not?

Did you want to,follow up your reading (how?) with researching some of these topics yourself?

What most surprised or intrigued you in this book?

About the 'Hearing Others' Voices' series

The series *(not* exam or curriculum related) aims to provide young adults with accessibly written introductions to topics, insights, and voices beyond their usual experience.

Hearing Others' Voices (https://www.balestier.com/category/hearing-others-voices/) is a transcultural and transdisciplinary book series edited by British anthropologist Ruth Finnegan and Taiwanese physicist Roh-Suan Tung. The books are designed, in straightforward language, to enlighten and attract general readers, undergraduates, young adults and, above all, teenagers about recent advances in thought, overlooked areas of the world, and key issues of the day.

They tackle perpetual topics of interest such as mental health, the nature of the universe, of music and of storms, shamanism, voices of the Christian west, and our awesome minds and bodies. They provide introductions, accessibly written (often illustrated, and linked to audio-video internet material), to the latest insights into fascinating and perpetual topics such as the different ways pain has been handled in the past and present, the history of children, the nature of poetry, and more. They attempt, too, to reveal something of the wisdom of the often dismissed traditions of, for example, hunters and gatherers, Amazonia, native Americans, and Africa, bringing a deeper and more balanced perspective on who we are, and how the world has come to be as it is.

The first volumes to be released (preceded by the October launch of the Chinese version of Rob Janoff's amazing personal account) were:

Taking a bite out off the apple, a graphic designer's tale, Rob Janoff, the world's most famous logo designer

Time for the world to learn from Africa, Ruth Finnegan, anthropologist,

Voices from the Christian west, Venerable Archdeacon Arthur Hawes, theologian, Emeritus Archdeacon, Lincoln Cathedral

From blank canvas to garment: a creative journey of discovery, Marella Campagna, fashion designer

Listen world! Evelyn Glennie, the workf's premier solo percussionist

Voices from Native American knowledge systems: something to learn, Clara Sue Kidsell, Emeritus Professor, University of North Carolina

For peace. Voices against the fog and blood of war, edited by Ruth Finnegan, anthropologist

. Other volumes (2019) include

Grass: miracle from the earth, David Campbell Callender, Irish naturalist

Decisions, decisions, decisions, with compassion and love: voice from the headteacher's study, A headteacher

Being a poet, being an Archbishop, being human, Rowan Williams, writer, poet, theologian, previously Archbishop of Canterbury

Further books will appear over the coming years, many also with translations into Chinese and other world languages.

Join our community

https://m.facebook.com/HearingOthers/

9 781911 221579